# THE CHURCHWARDEN'S YEAR

# THE CHURCHWARDEN'S YEAR

## A Calendar of Church Maintenance

*Illustrated by Graham Jeffery*


**CHURCH HOUSE PUBLISHING**
Church House, Great Smith Street, London SW1P 3NZ

**ISBN 0 7151 7555 6**

Published for the Council for the Care of Churches by Church House Publishing 1989
Second Impression 1992



With thanks to J.H. Edelston and others.

Printed in England by Rapier Press Ltd.

*O God, by whose spirit*
*the whole church is governed*
*and sanctified, grant to*
*us we beseech thee, that*
*by that same spirit we*
*may love thee and serve*
*thee.* Through Jesus Christ
Our Lord.

1. Stow away Christmas decorations.

2. Check your boiler and make sure that the frost thermostat is in good working order.

3. Be sure your rainwater gutters (especially valley gutters), hopper heads, downspouts, gulleys and drains are clean and working satisfactorily.
(The best time to do this is when it is raining!)

4. Are you certain that you have protected any exposed water tanks, your water pipes, heating pipes and oil-feed pipe against severe frost?

Frost Report

1. Make sure that no slates have slipped during snow.

2. Check that none of the gutters and downspouts has been damaged by snow and frost.

3. Organise spring-cleaning for April (a working party?)

Calendar
EPIPHANY. Pray for Gentiles, & check Boiler
LENT. Fast, Lament & OIL hinges
EASTER. Rejoice! and Clear out Jackdaws & nesting Pigeons

1. Look carefully at the roofs for frost, snow and wind damage (easiest with a pair of binoculars!).

2. Check all stonework and brickwork for frost damage.

3. Check all types of gutters and downspouts for storm damage.

4. Can any damage caused be claimed on your insurance? (Is your cover up-to-date?) If you can't find the policy, ask the insurers for a duplicate!

5. Oil the hinges and locks to your doors and windows.

6. Not long to the Annual Parish Meeting and your progress report on necessary repairs to the fabric! Check on the last quinquennial report (see your architect perhaps?)

Wall
Safe

1. Change of Churchwarden at the Annual Meeting? Check the Inventory.

2. Clean out gutters and downspouts.

3. Clear pigeons, jackdaws and others out of the roofs, tower and spire and birdproof them.

4. Have a good spring-clean in the church, tower, etc. (see February!)

5. Ask the bell captain if the bells and ringing chamber are in good order. Check the condition of any ladders in the tower.

6. Clean up the churchyard.

7. Start cutting the grass.

8. Archdeacon's visitation soon? Make sure you know where the Inventory is, and that it is up-to-date!

The blackbird hath found
herself an house, and
the Jackdaw a nest
Where she may lay
her young
even thy belfry
O Lord of Hosts

1. Shut down the heating system, have the boiler serviced and leave boiler house and boiler well ventilated to prevent condensation.

2. Get the electrics checked, especially those of the heating system.

3. Clear gutters, downspouts and roof space again.

4. Check security arrangements – who holds which keys and where are they kept?

5. Clear the vegetation from around the outside walls of the church.

6. Chase those pigeons and jackdaws out again!

1. Check that windows which open are in good working order. Ventilate church on dry days.

2. Look out for woodworm or death watch beetle – This is the month when the larvae hatch and the beetles fly.

3. Has your lightning conductor been checked during the last five years?

4. Tidy up the churchyard and cut the grass.

10
9
Trinity
8

1. Look out for fungus and dry rot.

2. Check any bird screens.

3. Take a good look at the notice board: is it in good order and tidy? Are the notices well-presented and up-to-date?

4. Would draught-proofing doors and windows save fuel next winter? (Remember to get the necessary permissions and consult your architect if necessary!) Any double-glazing ought to be professional.

*Go placidly among the water spouts and form-filling*
*And remember what peace may be had at* Matins.
*There will always be, in any diocese,*
*churches better organised than your own.*
Also, *it may be,*
*parishes where spires are falling down*
*and where the boiler has stopped working*
*altogether.*

But *you will know, in the providence of* God,
*that such parishes have a special part*
*to play*
*in the kingdom.*

Be*yond, then, a gentle disciplining of yourself*
*and a checking of the tiles in autumn*
*and of all things at the* Quinquennial,
Do *not be over zealous*
*in the checking of the water-pipes.*

Ful*fil, as best you may,*
*all the burdens and difficulties*
*of your* Wardenhood.

A*nd remember,*
*for all the coldness of the church at* Christmas,
It *is still* God'*s church.*
A*nd for all the faculties and administrations*
*you endure,*
Y*ou are still his child.*

(*with apologies to* 'Desiderata', *by* Max Ehrmann)

Have a holiday.

*'Lighten our darkness . . .*

1. Replace any broken bulbs (don't forget outside lights).
2. Test the boiler and check the heating system for leaks. Remember to bleed the radiators.
3. Check where necessary that you have adequate solid fuel for the winter.
4. Has the organ been tuned? Was the wiring to the blower and console light checked? Make sure no rubbish is accumulating behind the organ.
5. Keep mowing the grass!

Anne

1. Creosote the snowboards and repair any which have broken. Put them in place.

2. See that any exposed water tanks, water pipes, heating pipes and the oil-feed pipe are protected ready against frost.

3. Clean out autumn leaves from the gutters, downspouts and drains.

4. Watch out for an invasion of mice!

5. Make sure Harvest Festival decorations are cleared away.

6. Clear up the churchyard for the winter.

7. Service the mowing machine.

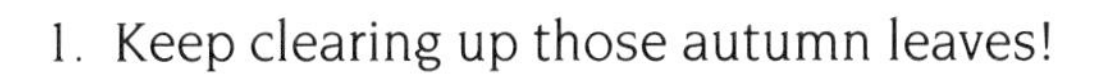

1. Keep clearing up those autumn leaves!

2. Check that the flagpole is secure.

3. Keep your church warm and dry. It will pay you well in the long term.

Sermon
Notes

1. Check frost protection.

2. Be ready for snow. It sometimes comes early.

3. See that all repairs or alterations made during the year have been entered in the Church Log Book (from SPCK or Church House Bookshop).

4. Clean church and decorate for Christmas.

5. Ensure fire-extinguishers have been serviced during the year – remember those candle-lit seasonal events!

6. Have a happy Christmas!

John Smith
Churchwarden
1824-1882
Peace at last